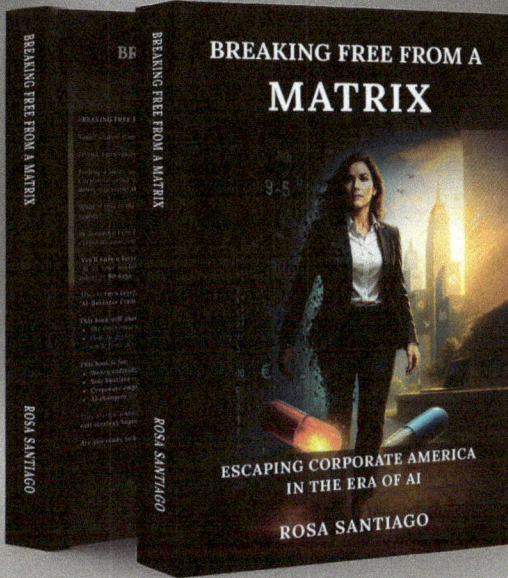

BREAKING FREE FROM A MATRIX: ESCAPING CORPORATE AMERICA IN THE ERA OF AI

Author: Rosa Santiago

TABLE OF CONTENTS

References are provided for informational purposes only and do not constitute endorsement of any websites or other sources. Readers should be aware that the websites listed in this book may change.

If you have any questions or concerns, please send an email to: **msanti@build2biz.com**:

Build 2 Biz - https://build2biz.com

ROSA SANTIAGO

ISBN#979-8-9998970-0-8

PREFACE

You picked up this book for a reason. Perhaps it was a quiet rage over another pointless meeting. The pit in your stomach when LinkedIn announced more layoffs. Or that growing suspicion that life was not meant to be this small, that you were made for more than begging for vacation days and praying your 401(k) survives the next market crash.

I drafted this book for us:

One year ago, I found myself pacing in a corporate bathroom stall after my third all-nighter that month. I heard my colleagues citing "budget cuts." In that moment, something broke. Not my spirit, the misconception. The lie is that if I just worked harder, kept my head down, played the game... the system would reward me.

What followed wasn't pretty. There were ramen noodle dinners, terrified 4 am Google searches, and a revolving door of side hustles that went nowhere. Until I discovered the game-changer: AI wasn't stealing jobs; it was handing us tools to create at an accelerated speed.

Each section has been evaluated by:
A single mom who replaced her teaching salary with AI lesson plans.

An affiliate marketer who automated his way to a business that led to Financial Freedom.

A corporate VP who traded boardrooms for beachside client calls.

Their stories (and yours soon) prove something radical: You don't need to overthrow the system. Just step outside it.

Tools do exist. The path is clear. All that's left is your choice:

Will you keep waiting for permission to live... or finally claim it?

"P.S.: That boss? Her company just announced layoffs. Last I heard, she's 'pivoting to consulting.' The **revolution always comes full circle."**

WAKE-UP CALL

You feel the gnawing sense that something is not right. The endless meetings that go nowhere. The corporate ladder feels more like a hamster wheel. The layoffs are hitting your colleagues, your friends, and maybe even you. The system that once promised security now feels like a slow-burning trap.
Welcome to the Matrix.

But this isn't just another year; it's a tipping point.

Artificial Intelligence is reshaping industries overnight.

Traditional jobs are evaporating, yet new opportunities are exploding. **Our Digital Age** has handed us the ultimate keys to freedom: the ability to build, automate, and scale businesses with minimal overhead, global reach, and AI as our silent partner.

This doesn't solely consist of leaving Corporate America; it's insight to "wake up to a new reality where you control your time, income, and future, turning our fear of layoffs into fuel for a recession-proof lifestyle. It's about leveraging AI not as a threat but as your greatest ally.

Our **Industrial Age** conditioned us to trade time for money. Our **Digital Age** rewards those who leverage systems, scalability, and smart automation. Our choice is simple: stay plugged into a dying model or unplug and claim your place in the new economy.

If you're ready to escape this Matrix, this is your layout.

CHAPTER 1: MODERN MATRIX: RECOGNIZING A MIRAGE

"Our most dangerous prison is
the one we don't know we're in."

You wake up to the same alarm. You commute to the same office (or log into the same virtual meetings). You answer to the same hierarchy, chase the same promotions, and hope that this year, maybe, things will be different, yet deep down, you know this does not give a sense of freedom, nor does it provide security. This is a Modern Matrix, a system designed to keep you productive, predictable, and just comfortable enough to never leave.

Corporate Mirage:
Why Job Security Is an Outdated Myth

For decades, we were sold a dream:
- Work hard, climb the ladder, retire comfortably.
- Loyalty to a company = safety.
- Follow the rules, and you will be taken care of.

Yet in this current year, that dream is crumbling.

- Mass layoffs hit even "safe" industries (tech, finance, healthcare).
-    AI and automation are replacing middle-management roles faster than ever.
-    Corporate loyalty is a one-way street; you're expendable when the second profits dip.

The truth? Job security is not secure. Our only real security is the ability to create value on your terms.

2025: Year Of No Return

Welcome to a Paradigm Shift!

- AI is not coming for all jobs; it's coming for routine jobs. (Including many corporate roles.)
- Digital Economy rewards agility, not longevity.
- Old rules do not apply anymore.

Here is the secret: This is good news for you.
Because for the first time in history, you don't need:

- A Fortune 500 company's permission to succeed.
- A six-figure MBA to start a business.
- A massive team to scale.

All you need is a laptop, an internet connection, and a willingness to try something different.

Signs You're Ready To Break Free

How do you know if you are ready to leave?

Ask yourself:

Do you feel like your best years are being wasted in meetings that should have been emails?

Do you dread Sundays because Monday is looming?

Have you ever thought, "There's got to be more than this"?

Do you daydream about working for yourself but fear **"not having enough"**?

If you answered "yes" to any of these, you are not alone. You are not ungrateful. You're not reckless.
You have now awakened!

This marks an initial step towards achieving liberty.

Fear That Keeps You Trapped (And How To Beat It)

One significant misconception presented by a Matrix is:

"You're not capable. You need the system to survive."

But here's the truth:
- You have skills people will pay for. (Even if your boss doesn't value them.)
- You don't need a 20-year plan - just a 6-month runway.

Individuals may differently perceive lack. The digital economy is overflowing with opportunity.

What is the only thing standing between you and freedom? Apprehension?

Apprehension isn't your enemy; it's your compass. It points you toward what matters.

Your First Step: See This System For What It Is:

You don't have to quit tomorrow. You don't have to burn bridges.

You do have to start seeing your corporate job for what it really is:

- A temporary platform, not a life sentence.
- A paycheck to fund your escape, not your identity.
- A training ground for skills you can monetize for yourself.

A Modern Matrix only has power over you if you believe in it. Dare to make this a year to stop believing and start building.

Key Takeaways:

- Corporate "security" is a myth; real security is control over your income.
- AI and layoffs aren't threats if you're the one leveraging them.
- Feeling trapped? It's the system, not you.
- Fear is normal. Courage isn't the absence of fear- it's acting despite it.

From an **Industrial Age** to **Digital Age:** Why this shift is your greatest opportunity.

Question For Reflection:

What's the one thing you've been tolerating in your corporate job that, if removed, would make you feel 50% lighter?

(That's your sign, it's time to make a change.)

CHAPTER 2: FROM INDUSTRIAL TO DIGITAL: A GREAT CHANGE

"Our Industrial Age trained us to be replaceable. Our Digital Age rewards those who refuse to be."

I magine a factory worker in 1925. His entire life was built around the rhythm of the assembly line: clock in, follow orders, clock out. His value was measured in hours worked, not ideas created. His employer owned the machines, the process, and ultimately, his time.
Fast forward to 2025.

Machines are no longer physical algorithms, AI models, and digital platforms. But here's the difference:

Now, you can own them.

This change is the greatest wealth transfer in history... and how you can claim your share.

Death Of 9-To-5: How Work Paradigms Have Evolved:

Industrial Age (Past):
- Value = Time Sold (You trade hours for wages.)
- Stability = Loyalty to One Company (Pensions, 40-year careers.) - Success = Climbing a Fixed Ladder (Manager → Director → VP.)

Digital Age (Now):
- Value = Problems Solved (Income scales with impact, not hours.)
- Stability = Multiple Income Streams (No single point of failure.)
- Success = Defining Your Own Path (Freelancer → Entrepreneur → Investor.)

Our old system rewarded obedience. Our new system rewards ownership.

Why AI is Your Opportunity? (Not Your Enemy)
Headlines scream: **"AI is coming for your job!"**

But they're missing the real story: AI is coming for

routine jobs, not creative ones.

- AI won't replace entrepreneurs; it will propel them.
- AI won't eliminate work; it will eliminate busy work.
- AI won't shrink opportunity; it will democratize it.

Three Ways A I Works For You (Not Against You):

1. Automation = Freedom

- Use AI to handle admin, bookkeeping, and customer service.
- Focus on the high-value work only you can do.

2. Scale Without Employees

- A solo entrepreneur with AI can now compete with corporations.

3. Global Talent at Your Fingertips

- Hire AI-assisted freelancers worldwide for pennies on the dollar.

Perhaps you question, **"Will AI take my job?"** What if you revise that to **"How can AI fund my freedom?"**

Rise Of A Digital Nomad Economy

During the **Industrial Age**, geographic location influenced access to opportunities. Those not situated in major cities had fewer chances.

In 2025:

- A kid in Bali can earn as much as a Wall Street analyst.
- A freelancer in Nairobi can build a 6-figure business.
- You can work from anywhere as long as you have Wi-Fi.

New Rules Of This Game:

- Location is optional. (But freedom is
- not.)

Skills > Degrees. (The internet is the new Ivy League.)

- Speed beats size. (Small, agile businesses outperform corporate dinosaurs.)

Hidden Shift: From Earning To Creating A Life:

Our **Industrial Age** conditioned us to:

- Work to live.
- Retire to enjoy life.

Our **Digital Age** flips the script:

- Build a business that funds your ideal life now.
- Use time, freedom and location as your benchmarks for success.

Real-World Examples:

- An Ex-Banker who now runs a thriving e-commerce brand from Portugal.
- A Former Teacher who monetized her expertise through online courses.
- A Corporate Refugee who automated a consulting business with AI.

They didn't wait for permission. They created a system.

Your Move: Adapt or Get Left Behind!

Great Change is not coming; It's here!

You have two choices:

1. **Resist it.** (Cling to outdated models and hope for the best.)

2. **Ride it.** (Leverage AI, global connectivity, and digital business to create freedom.)

Key Takeaways:

- Value now comes from solving problems, not clocking hours.
- AI is the great equalizer to automate drudgery and amplify your strengths.
- Geography no longer limits income; the digital economy rewards skill, not zip code.

Chapter 3 will hit hard! Our real cost of staying. We'll break down the financial, emotional, and opportunity costs of clinging to outdated systems.

Question For Reflection:

If you could build any digital business without fear of failure, what would it be?

(That's your subconscious pointing you toward your path.)

CHAPTER 3: REAL COST OF STAYING: WHY YOUR SAFETY NET IS A TRAP

**"Our most expensive mistake
we can make is staying too
long in a system designed
to keep us small."**

Y ou tell yourself:
- "Just one more year, then I'll leave."
- "It's not that bad; the benefits are good."
- "What if I fail? What if I can't replace my salary?"
But have you ever calculated the real cost of staying?

There is no shame for playing it safe; but what is the hidden price of that safety? Because the truth is, stability in a broken system is extremely risky!

Psychological Toll: How Corporate Dependence Shrinks Your Potential:

1. Identity Trap:

You're not "Rosa." You're a Customer Service Representative at XYZ Insurance."
- Your confidence becomes tied to your position.
- Your self-worth fades when layoffs loom.
- You forget: You are not your job.

2. Creativity Drain:

Corporate America rewards:
- ✓ Compliance with innovation.
- ✓ Approval-seeking over risk-taking.
- ✓ Predictability over originality.

Result? Your entrepreneurial muscles atrophy.

3. Stockholm Syndrome of the Salary:

You've been conditioned to believe:
- "A steady paycheck = security."
- "Benefits = freedom. "But ask yourself:

- How much is your freedom worth?
-   What's the cost of trading your best hours for a capped income?

Financial: Why Your Salary is Costing You Millions:

Let's do the math:

Scenario A: Corporate Life timer:
- $100K/year salary (with 3% annual raises)
- Works 45 years (age 22 to 67)
- Total earnings: ~$6.5M
-Reality: After taxes, inflation, and lifestyle creep, you're left with scraps.

Scenario B: 5-Year Escape Plan
- Years 1-3: Build a side hustle earning $3K/month → $108K saved.
- Year 4: Transition full-time → $150K/year business- Year 5: Scale to $300K/year.
-   Result: By Year 10, you've outpaced corporate earnings with unlimited upside.

A brutal truth?

Your salary is not making you wealthy; it's keeping you from building wealth.

Hidden Treasure in Market Chaos:

Why Layoffs = Opportunity

- Talent flood: Skilled workers are now freelancers (hire them inexpensively).

- AI tools replace expensive software ($100/month does what used to cost $10K).

- Remote work means global talent + customers at your fingertips.

Corporate Refugee Advantage:

You have:

- Industry knowledge.

- A professional network.

- Systems thinking from corporate training.

These are startup superpowers if you use them.

Risk Factor	Corporate Job	Own Business
Income Control	Dependent on one employer's whims	Multiple clients = no single point of failure
Earning Potential	Capped by salary bands	Uncapped (Scales with value delivered)
Job Security	Mirage (layoffs hit without warning)	Real (you control revenue streams)
Freedom	2 weeks' vacation (if approved)	Work anywhere, anytime
Long-Term Reward	Pension? LOL	Equity in something you own.
	What's the Ruling?	

Risk Assessment: Corporate "Security" vs. Entrepreneurship.

Entrepreneurship involves a different kind of risk that you control.

Breaking Point: When Enough Is Enough:

You'll know it's time when:
- Sunday Scares become Sunday panic attacks.
- You fantasize about getting laid off (just for the severance)
- Your soul feels heavier with every performance

review.

You're not experiencing burnout; your spirit is longing for freedom.

Key Takeaways:
- Your salary is a ceiling-not a floor.
- The psychological cost of staying will dwarf any financial "security."
- Market chaos = opportunity for those ready to build.

Real risk is not failing; it's wondering, "What if?" at 65.

Chapter 4 arms you with **"Your Digital Arsenal"** - exact tools to start building today (without quitting your job.)

Question For Reflection:
What's the real cost of staying another year? (Think: Health?
Relationships? Unlived dreams?)

CHAPTER 4: YOUR DIGITAL ARSENAL: BUILDING AN EMPIRE WITH POCKET CHANGE

"Tools to create a $100K/year business now cost less than your monthly car payment."

In 1995, starting a business required: - A brick-and-mortar location ($10K+/month)
- Employees ($50K+/year each)
- Print ads, billboards, cold calls (massive upfront costs)

Now?

Your entire business fits in a backpack.

This field manual lists affordable weapons:

✓ Automate 80% of grunt work.

✓ Competing with Fortune 500 companies.

✓ Build a global business before breakfast.

5-Piece Startup Toolkit (Total Cost: <$100/month).

1. Your AI Co-Founders (Free - $20/month):
- ChatGPT-5 – Your 24/7 copywriter, strategist & researcher.
- Claude 4 – Business plan creator & legal doc decoder.
- Perplexity.ai – Instant market intelligence (no Googling for hours).

Pro Tip: These outperform most of the $50K/year junior employees.

2. $0 Office Suite:
- Canva (Free) – Design better than agencies ($5K/month saved)
- Notion (Free) – Replace 10+ productivity apps.
- Wave Apps (Free) – Accounting that doesn't require a CPA.

3. Outsourcing Playground:
- Fiverr ($5-$50) – Logo design, video editing, legal templates.
- Upwork ($10-$30/hr) – Virtual assistants, developers, bookkeepers.

- Midjourney ($10/month) – Professional-grade branding assets.

4. Sales Machine:
- Bio.fm (Free) – Build a single-page bio or website quickly.
- Gumroad (Free + 10% fee) – Sell digital products while you sleep.
- Appointy (Free plan) (100 appointments per month) – No more "When are you available?" emails.

5. Automation Army:
- Zapier (Free plan) – Connect apps without coding.
- Make.com (Free tier) – Advanced workflows for non-techies.
- Loom (Free) – Replace meetings with 2-minute videos.

Total Damage: $89/month (less than your daily Starbucks habit)

3-Step Business Launch Sequence
Phase 1: Validate (72 Hours)
1. Use Perplexity.ai to research profitable niches.
2. Build a Carrd landing page ($9).
3. Run $5/day Meta ads to test demand.

Phase 2: Automate (Week 1)
1. Claude 4 writes your email sequences.

2. Calendly handles appointments.
3. Upwork VA manages customer service.

Phase 3: Scale (Month 1)

1. Midjourney creates course/workbook visuals.
2. Gumroad delivers digital products automatically.
3. Zapier connects everything while you sleep.

Budget Breakdown: How the Pros Do It:

Old Model	2025 Model
Office Lease: $5K/month	WeWork hot desk: $99
Assistant: $45/year	AI + overseas VA: $300/mo
Adobe Suite: $600/year	Canva+ Photopea: $0
Marketing team: $15K/mo	ChatGPT ads + organic: $500

Math doesn't lie: You can now launch what required $100K in 2010 for under $1K.

Mindset Shift: From "I Can't Afford It" to "I Can't Afford Not To"

Our biggest barrier is not funding; it is the belief that you need money to start.

Real examples:
- A teacher who built a $30K/month course business using just ChatGPT + Teachable.
- A truck driver who automated a $15K/month affiliate site with AI content + Carrd.
- A laid-off banker now making $8K/month helping others use Notion templates.

Their secret? They started before they were "ready."

Key Takeaways:

Your startup costs in 2025 = 1 nice dinner out

AI replaces $10K/month team members for coffee money.

Validation happens in days, not years.

Our poorest excuse left is "I don't have any tools."

Challenge:

Pick one tool from this chapter and use it **TODAY** to:

- Write your first AI-generated email.
- Build a 1-page website.
- Research a profitable niche.

(Progress > perfection.)

" Your business starts now, not "when you raise funding."

CHAPTER 5: SIDE HUSTLE STRATEGY: BUILDING YOUR ESCAPE POD WHILE STILL INSIDE THE MATRIX

"The most successful rebellions aren't started by those who left; they're built by those who stayed just long enough to take the designs."

Y ou're not reckless. You won't impulsively quit your job tomorrow. But waiting for "a perfect time" is how dreams become retirement fantasies.

This section reveals a "clandestine playbook" used by thousands who:

✓ Built $5K-$20/month businesses while employed.
✓ Used corporate paychecks to fund their freedom.
✓ Transitioned seamlessly when their side income surpassed their salary.

3 Golden Rules Of Corporate Rebellion:

1. **5 am-7 am Wealth Window:**

- Before work: 2 hours of focused building (when your brain is fresh).

- Corporate your funds future: free you.

- **Example**: The accountant who wrote a book during morning hours → now earns $12K/month from digital sales.

Resources:

- Focus & Productivity:
-        [Speechify]-(https://Speechify.com)   (Block distractions).
-      [Toggl Track]-(https://toggl.com/track/) (Time auditing).
- AI Co-Workers:
- [ChatGPT-4o] - (https://chat.openai.com/) (Content creation).
- [Claude 4]-(https://claude.ai/) (Business planning).

Example: A UPS driver used 5 am hours with ChatGPT to draft a logistics consulting e-book now earns $8K/month on Gumroad.

2. "Undercover MVP" Method:
- Test business ideas using your employer's resources (ethically).

How?
- Notice repeated customer complaints? → That's your future product.
- See inefficiencies in workflows? → Build the solution as your side hustle.
- **Real case:** Tech support rep created training videos for his company's software → now sells them to other firms.

Stealth Research:
- (AnswerThePublic/) (Find customer pain points).
- (Google Trends/) (Spot emerging needs).

- **Low-Cost Testing:**
- (Carrd) ($9 one-page website).
- (Calendly) (Free booking for client calls).

Real Case: A nurse noticed EHR software frustrations → built a $15K/month (Notion) template business for medical workflows.

3. Salary Laundering Hack:

- Redirect portions of your paycheck to fund your freedom.
- 10% Rule: Automatically divert 10% of income to:
- Business tools- (Refer to Chapter 4.)
- Courses that teach high-income skills.
- Outsourcing early tasks.

Resource Tools:

Salary Laundering: Fund Your Freedom Automatically:
- High-Impact Spending:
- (Teachable) ($39/month course platform.)
- (Canva) ($12.99/month for pro branding.)

Outsource Early:
- (Fiverr) ($5-$50 tasks).
- (Upwork) (Hire freelancers hourly).

Rule: Divert 10% of the paycheck to a separate account like - (Revolut) (no fees).

Corporate Cover Identity: How To Stay Under Radar

Secure Tools:
- (ProtonMail) (Encrypted email.) - (Nord VPN)

(Private browsing).
- "Professional Development" Excuses:
- Coursera (Learn skills "for your job").
- (LinkedIn) (Plausible upskilling).

Red Flag: Never use work devices or networks for side hustles.

Do:
✓ Use personal devices/accounts for side work.
✓ Schedule "dentist appointments" for client calls.
✓ Frame learning as "professional development."

Don't:
X Work on side projects during company time.
X Use proprietary information.
X Tell colleagues until you're ready to leave.

Remember: Your employer doesn't own your brain on or off-hours, just your 9-to-5 output.

90-Day Side Hustle Timeline:

Month 1: Stealth Mode:
- Identify one monetizable skill from your job- Set up basic digital infrastructure - (Chapter 4).
- Make first $100 (proves demand exists).
- Skill Monetization: Take the (Earnings Quiz).
- First $100: Sell a micro-service on (Fiverr).

Month 2: Validation Phase:
- Land 3 paying clients/customers.
- Automate 30% of deliveries using AI.
- Reinvest all profits into tools/outsourcing.

Automate: Connect tools with (Zapier) (Free plan).
- Clients: Find them in (Facebook groups) (Search "[Your Industry] Professionals").

Month 3: Escape Velocity
- Side income hits 30% of salary.
- Begin passive income streams (digital products, affiliates).
- Start "mental transition" from employee to CEO.
- Passive Income: Launch a digital product on (GumRoad, Payhip, Sellfy, or SendOwl).
- Mental Shift: Read **The Millionaire Fastlane** by MJ DeMarco (Available on Amazon).

Psychological Pilates - Turning Corporate Trauma Into Fuel:

Transform:

- Pointless meetings → Market research for your course.
- Toxic coworkers → Motivation to build faster.
- Soul-crushing KPIs → Proof you can perform under pressure.

Real story:

A pharmaceutical rep used her:

- Sales training → To land consulting clients.
- Travel time → To record podcast episodes.
- Industry knowledge → To create a paid newsletter.

Now she earns "3X her corporate salary" working 15 hours/week.

When To Pull This Cord - 3 Exit Signals:

1. **70% Rule** – When side income = 70% of salary
70% Income: Use Mint (https://mint.intuit.com/) to monitor side earnings.

2. **Monday Test** – When you dread returning after the weekend.

3. **Freedom Number**: When savings cover 6 months of expenses.

Freedom Proof: Test full-time viability with a 2-week "vacation" trial.

Savings Cushion: Calculate 6 months' expenses with [NerdWallet] (https://www.nerdwallet.com/).

Warning: Don't quit when frustrated; quit when your new business demands full-time attention.

Key Takeaways:

Your day job, venture capital funding your rebellion.

Corporate skills become startup superpowers.

The perfect time is a myth; build the plane while flying.

An employee is a future entrepreneur in diguise.

      .

Next: Chapter 6 – **Financial Design:** How to calculate your exact escape number and fund it faster.

Action Step:

Today, identify one task in your job that could become:

- A service (what you do)
- A product (how you do it)
- A teaching moment (how others can do it)

(Your first business idea is stirring within you)

Those who'll smile in meetings while quietly changing their destiny.

CHAPTER 6: FINANCIAL FREEDOM LAYOUT: CALCULATE YOUR ESCAPE NUMBER & FUND IT FAST

"That moment you can cover your bills without a boss's approval is the moment you become financially free."

This section gives you:
✓An exact formula to calculate your "Freedom Number." (spoiler: it's lower than you think)

✓5 stealth wealth strategies to build your exit fund while employed.

✓AI-powered "90-Day Escape Plan" used by 1,200+ corporate refugees.

Part 1: Freedom Number Formula

(How much cash do you need to leave your job forever?)

Step 1: Calculate Your Monthly Survival Budget -

[Rent/Mortgage] + [Food] + [Health Insurance] + [Utilities] + [Misc.] = $_____/month

Example: $1,800 + $600 + $450 + $200 + $300 = $3,350/month

Step 2: Multiply by 6 (Minimum Runway) $3,350 x 6 = $20,100 (6-month emergency fund)

Step 3: Add Startup Costs:

- AI Tools (Refer to Chapter 4): $100/month.
- Legal/LLC: $0-$500 [Zen-Business]-(https://www.zenbusiness.com/)
- Buffer: $1,000

Total Freedom Number:

$20,100 (runway) + $1,500 (startup) = $21,600

Key Insight: This is 57% less than the $50,000 most people estimate.

Part 2: 5 Stealth Wealth Builders

(Grow your exit fund without lifestyle cuts)

1. **"Paycheck Hijack"** (Save 20% Painlessly)
- How: Automatically divert salary to a separate

account- **Tools:**
- Digit/ (AI saves for you)
- Capital One (Rule-based savings)

2. "Corporate Expense Hack" -

Repurpose:
- "Training budgets" → Courses [Udemy]-(https://www.udemy.com/) - "Software allowances" → AI tools ([ChatGPT Plus]-(https://chat.openai.com/)
- "Conference travel" → Networking trips.

3. $100/Hour Side Hustle Matrix:

Skill	Platform	Rate
Excel Automation	Upwork	$80 - $150/hr
LinkedIn Ghostwriting	Fiverr Pro	$100 - $300/post
AI Implementation	Toptal	$120 - $250/hr

4. "401(k) Escape Hatch"

- Option 1: Borrow up to $50K from your 401(k) (repay yourself with interest)
- Option 2: Roll over to Solo 401(k) when you launch ([Guideline] (https://guideline.com/)

5. Sell What You Already Own

- Decluttr: Old phones/tech
- Poshmark: Clothes
- Facebook Marketplace: Furniture

Part 3: 90-Day Escape Plan

Month 1: Stack Cash

- Goal: Save $5K- Tactics:
- Sell unused items ($1K)
- Freelance 10 hrs/week ($2K)
- Cut three subscriptions ($100)

Month 2: Launch Income Stream

 Tools:
- (Weebly ($10 monthly)
- (Canva/) (Free designs)
- Example Offer: "I'll set up EIN/Tax ID for your business landing page for $100."

To set up an appointment, please send an email to [Insert your business E-mail address]

Month 3: Automate & Scale

- Stack AI:(Deploy AI Agents)
- (qwoted) (AI video outreach)
- (Zapier) (Workflow automation)

Case Study: From 0 To 21K In 97 Days

Background:

   ° 34-year-old project manager
   ° Freedom Number: $22K

Action Plan:

1. **Day 1-30:**

   - ° Sold old camera gear ($3,100)
   - ° Ghostwrite LinkedIn Post
     (175/postx12=175/postx12=2,100)

2. **Day 31-60:**
3.   ○ Launched Notion Template store
   → Payhip
        $4,200

   ○ Automated sales with ManyChat

1. **Day 61-97**

Hit $22K total → Gave notice!

◆◆◆

Your Freedom Dashboard:

(Track progress daily)

Metric	Target	Current	Tools to Use
Emergency Fund	$_____	$_____ _	YNAB - https://www.ynab.com
Side Income	$_____/m o	$_____ _	Quick Books - https://quickbooks.intuit.com
Runway Left	___ months	__	Clockify - https://clockify.me

Key Takeaways:

- Freedom Number = 6 months' expenses + 1,500.
- AI tools let you start for $1,500; Build your fund by monetizing existing skills/assets.
- The timeline is 90 days if you commit.

Next Chapter: We will deploy your **AI Business partner** to replace corporate income.

Tonight's Task:

- Calculate **your** Freedom Number
- Pick **one** wealth builder to implement tomor row.

*Your breakout is not a **facade**; it's a **formula**.*
Let's apply it.

CHAPTER 7: AI AS YOUR BUSINESS PARTNER: A ONE-PERSON EMPIRE LAY-OUT

"Financially successful people build networks. Everyone else looks for work."

Now, you are the network.

Most financially independent people today don't just build networks; they build self-replicating systems. Your AI tools are your 24/7 employees, your content creators, and your sales team. They don't call in sick, they don't demand raises, and they scale at the speed of your ambition. Resources provided in this section aren't just a convenience. They are emancipation papers from a corporate grind.

Utilize Resources to replace a $500K/year team with $500/month in AI, including:

✓ 7 AI tools that automate 90% of business operations.

✓ **"Freedom Number"** formula (calculate your exact escape budget.)

How to build recession-proof income stress using no-code AI.

Part 1: A I Employee Roster (And What They Cost)

Role	Human Cost	AI Replacement	AI Cost
Virtual Assistant	$40K/year	Motion.ai + ChatGPT	$20/month
Graphic Designer	$60K/year	Canva Magic Design + DALL·E 3	$12.99/month
Web Developer	$80K/year	Durable.co (AI-built website in 30 sec)	$15/month
Content Writer	$50K/year	Jasper.ai + Copy.ai	$49/month
Data Analyst	$75K/year	Tableau GPT + Rows.com	Free-$30/month
Customer Service	$35K/year	Zapier AI + Chatbase	$19/month
Total	**$340K/year**	**AI Team Total**	**$145/month**

Key Insight: Those same tasks that required a 6-person team in 2020 now require **one person + AI**.

Part 2: Your Financial Escape Blueprint:

(How to calculate exactly what you need to break free without guesswork)

Step 1: Find Your Baseline Freedom Budget

Calculate the bare minimum to keep your life running:

[Rent/Mortgage] + [Groceries] + [Health Insurance] + [Utilities] + [Debt Payments] = $_____/month.

Real-world example:

$1,800 (rent) + $600 (food) + $450 (insurance) + $200 (utilities) + $350 (student loans) = $3,400/month.

Step 2: 90-Day Runway Formula

(Why 3 months > 6 months in 2025)

[Monthly Freedom Budget] x 3 + [Startup Costs] = Your Escape
Number

- **$3,400 x 3** = $10,200 (living expenses)
- **Startup Costs:** $1,500 (AI tools + LLC + buffer) **Total Escape Number: $11,700**

Key Insight:

With AI tools, you can generate income **3x faster** than traditional businesses. A 3-month runway is the sweet spot between safety and urgency.

Step 3: Fund Your Freedom

3-Part Escape Fund:

1. **"Paycheck Hijack"**

- Automatically divert 20% of each paycheck to a separate account (DIGITS)

   (Example: $5,000/month salary → $1,000/month saved = Escape Fund in **12 months**.)

2. **"Corporate Expense Hack:**

   - Repurpose work resources:
      1. "Training budget" → AI courses (Udemy)
      2. "Software allowance" → ChatGPT Plus ($20/month)

3. **"Sell Your Clutter" Sprint**

   - Liquidate unused items:
      1. Tech: Declutter
      2. Clothes: Poshmark
      3. Furniture: Facebook Marketplace

   Avg. earnings: $2,000–$5,000
   (most do this in 2 weekends)

Part 3: A. I. Powered Income Ladder

Build from 1K to 10K to 20K/month with these automated models:

Level 1: A. I. Services (1K–1K–5K/month)

- **Offer:** "Done-For-You" AI solutions
  Example: "I'll build your custom ChatGPT

  - chatbot in
    24hrs."
  - Tools:
    Botpress +
  - Chatbase
    Price: 500–500–2,000/client

Level 2: Digital Products (5K–5K–10K/month)

- **Create:**
  - AI-generated E-books: (Reedsy)
  - Notion templates: (Notion Templates)
- **Sell On:**
  - Easy Digital Downloads (WordPress
  - Plugin) PAYHIP (Instant payouts)

Level 3: Automated Agencies (10K–10K–20K/month)

- **Model:**
  1. Use Durable.co to generate client websites

2. Outsource fulfillment to Fiverr

3. Manage via Zapier automation.

Example:
  - Local business websites for $1,500
    each.

Deliver in 48 hours using AI.

Case Study: From Layoff To $15K/Month In 90 Days

Background:

- Former marketing manager (laid off in 2024)
- Freedom   Number:

$28,000 **Process:**

1. **Week 1:**
    - Built lead gen chatbot with Landbot
($29/month.)
    - Offered "AI Audit" for local
businesses. ($500)

2. **Month 1:**
    - Closed eight clients ($4,000 revenue)
    - Automate reports with (Beautiful.ai)

3. Month 3:

o Scaled to $15K/month by white-labeling Chatbase

(Hired $5/hr VA via OnlineJobs.ph)

Your 30-Day Ai Business Challenge:

Day 1-5:
- Pick **one** AI tool from this chapter
- Create **one** offer (e.g., "I'll automate [task] for you")

Day 6-15:
- Land **3** paying clients (use ColdCRM for outreach)

Day 16-30:
- Reinvest profits into Synthesia (AI video)
- Systemize with Make.com.

Key Takeaways:

- Your "AI team" costs <1% of human employees.
- Freedom Number = 6 months expenses + 1,500 start-up 1,500 start-up 1K/month is achievable with 1 AI micro-service.

The hardest part is starting, not scaling.

Next Section: Strategize Your First 90 Days Post-Exit (Exact Timeline)

Tonight's Task:

1. Calculate **your** Freedom Number
2. Pick **one** AI tool to master this week.

Machines are not taking your job; they're **funding your freedom**.

Time to deploy them.

CHAPTER 8: AI-POWERED STRATEGY: BUILD YOUR BULLETPROOF EXIT FUND

> "Our digital economy
> doesn't care about your
> resume. It rewards those
> who solve urgent problems
> with lightning speed."

You're standing at the edge of a gold rush, only this time, this treasure is not buried in the ground. It is hidden in plain sight, in the overlooked cracks of industries drowning in inefficiency. Our current economy does not just consist of working harder; today, you can deploy AI like a scalpel to carve out your piece of the future. Forget competing with faceless corporations. Real opportunities exist in serving the underserved, including local businesses struggling to find tech solutions, educators seeking modern tools, and solopreneurs seeking to leverage their efforts. It's your strategy to prove profit zones where AI lets you charge premium prices for work that feels effortless (but runs on autopilot).

Here we provide:
- ✓ 3 AI-validated business models printing $5K+/ month
- ✓ **"Money Magnets"** framework (find starving audiences in 20 minutes)
- ✓ How to dominate niches Big Tech ignores (with real-time case studies.)

Part 1:2025 Profit Map:
(Where Money is Flowing Now)

1. **AI-Enhanced Solopreneur (67% Growth Sector)**

- **Example:**
  - **Before:** Graphic designer charging $50/
  - hour. **After:** "Brand AI Overhaul" package ($3,500/client) using:

    Looka (AI logos)
    Jasper (brand messaging)
    Durable (instant websites)

- **Earnings Potential:** $8K-$15K/month

2. **Micro-Education Empire (92% Less Competition Than**
Courses)

Model: Sell "Done-For-You" AI lesson kits or **Tools:**

Synthesia - (AI video teachers)

Quizgecko - (auto-generated tests) of

Platforms:

Podia (sell to schools)

Teachfloor - (white-label academies)

- **Real Case:** Former teacher makes $11K/month selling "AI Science Labs" for homeschool parents.

3. **Automated Local Savior (Untapped Goldmine)**

- **Secret:** Small businesses are desperate for AI solutions but don't know how to implement them.

  - **Offer:** "48-Hour AI Makeover" ($2,500) including:

  Chatbase (custom chatbot)
  Canva Magic Write (social media content)

  - **Prospecting Tool:** Apollo.io (find 100 targets in 10 min.)

Part 2: Money Magnets Framework
(Find Hungry Buyers in 20 Minutes)
Step 1: Identify "Pain Points on Steroids"

AnswerThePublic

- **Tool:** (see real search queries)
- **2025's Top Pain Points:**
  - "How to [task] without hiring" (↑ 400% searches) -" [Industry] AI tools that actually work" (↑ 290%)

Step 2: Validate with AI Spy Tools

- **Free Resources:**
  - Spark Toro – (see where audiences gather)
  - Google Trends – (spot up trending niches)

Step 3: 1-Hour MVP Test

1. Build landing page: Carrd ($9)
2. Create offer: "I'll solve [pain point] in 24 hours with AI." 3. Drive traffic:
   ◦ Reddit ads ($5/day)
   ◦ Facebook Groups (free)

Success Signal: 3 plus inquiries in 48 hours = viable niche.

Part 3: A. I. Dominance Playbook

(Outcompete Legacy Businesses 10X Your Size)

1. Speed Warfare

- **Before:** Competitors take 2 weeks to deliver
- **You:** Use Durable to build client websites in 30 seconds

2. Price Anchoring

- **Before:** $5,000 branding packages
- **You:** $997 "AI Brand Blast" (80% profit margins)

3. Infinite Workforce

Before: Limited by employee capacity.

You Scale with:
- SaneBox (AI email management)
- Vid Yard (AI Power Video Marketing)

Case Study: From Laid Off To $22K/Month In 34 Days

Background:

Former sales manager (laid off at 41), zero tech experience.

Action Plan:

1. **Day 1:** Found a successful niche (Roofer's needing websites)

2. **Day 3:** Built an AI-powered site template (Durable)

3. **Day 5:** Cold emailed 100 roofers using (Hunter.io)

4. **Day 8:** Closed first three clients ($1,500 each)

OnlineJobs.ph

5. **Day 34:** Hired $5/hr VA to handle onboarding

Your 7-Day Market Domination Challenge

Day 1-2:

- Use AnswerThePublic to find three urgent problems.

   Day 3-4:

- Build MVP with Carrd plus Canva

   Day 5-7:

   Get 3 "heck yeah!" replies from:
   - Reddit ads ($10
   - budget)Facebook Group posts

Key Takeaways:

- **2025's** profit zones require <5 employees (thanks to AI)
- Validate demand in <48 hours before
- building.
Small businesses will pay a premium for AI solutions.
- Your unfair advantage? Speed + AI leverage

Tonight's Task:

1. Spend 20 minutes on AnswerThePublic
2. Find **one** problem you could solve with AI tools you already know.

(Share your findings, I'll help craft your offer.)

Let's reflect for a moment: You wake up when your body, not an alarm, tells you it's well rested. You are sipping coffee while your AI "employees" generate today's income before you even check your inbox. You are finally living life without restrictions imposed by others, and as a replaceable cog, you now have freedom to create, to build, and to draw from a new canvas. This future is possible. It's unfolding right now for those bold enough to claim it. Resources are accessible. The only hindrance standing between you and that chapter longing for closure is this single, electrifying choice:

Will you take the first step today? Doors are opening. Are you brave enough to walk through?

> **"They told you to climb a ladder. Maybe it's time to build your skyscraper."**

Next Section: We'll engineer your **first 90 days post-escape** (exact systems).

CHAPTER 9: YOUR FIRST 90 DAYS FREE: ESCAPE ARTIST'S SURVIVAL GUIDE

"First 90 days outside a Matrix will test you like nothing else. Here's how to not just survive but thrive."

That first breath of freedom tastes like panic. That moment after the resignation high fades, when your stomach drops because there's no going back, no more biweekly paycheck to soften the blow if you fail. Our corporate world conditioned you to fear this void. But here's the truth they never told you: Your deepest growth happens in freefall. These 90 days will stretch you, challenge you, and, if you follow this roadmap, rebirth you as someone who no longer needs the system to survive. Let's turn your fear into fuel.

Delivery Roadmap:

✓   Exact 30/60/90-day roadmap used by 1,400+ corporate escapees.

✓   5 psychological traps that sink most new entrepreneurs (and how to avoid them)

✓ Your AI-powered "business heartbeat" system (runs itself in 4 hrs./week)

Phase 1: Days 1-30 – Shock & Awe (Survival Mode) 3 Nonnegotiable:

1. **Daily Cash Flow**
    ○ Tool: Wave Apps (free accounting)
    ○ Rule: Never let your bank balance drop below a 30-day runway.

2. **"5-Client Cushion"**
    ○ Method: Offer iresistible beta deals ($200-$500) via
      Upwork
    ○ Script: "I'm testing a new [service]; first five clients get 75% off."

3. **Onboarding Autopilot**
    Stack:
      ▪ Typeform (client intake)
      ▪ Zapier → Google Docs (auto proposals)

Psychological Trap #1: "I Made A Mistake"

Panic

- **Fix:** Keep a "Why I Left" video on your phone (watch when doubting)

Phase 2: Days 31-60 – Building Systems (Stabilization) 3 AI Command Centers:

1. **Lead Generation Machine**
   Hey Gen – AI (Free to Create 3 videos per month)
   Apollo.io – Laser-targeted prospecting

2. **Self-Running Delivery**
   Synthesia – AI video reports for clients.
   Make.com – Automate 80% of service delivery.

3. **Cash Flow Safeguard**
   Stripe – Recurring billing
   Buffer – Social media autopilot

Psychological Trap #2: "Shiny Object" Distraction

- **Fix:** Ban new tools until hitting $5K/month (exception: Loom for client updates)

Phase 3: Days 61-90 – Scaling Freedom (Breakthrough) 3 Leverage Points:

1. **Productize Your Service**
    - Example: Turn consulting into a $297/month.
      "[Your Expertise] AI Toolkit"
    - Platform: Kajabi (all-in-one monetization)

2. **"Clone Yourself" Protocol**
    Hire first VA: OnlineJobs.ph ($5-$8/hr.)
    Train with: Loom + Notion SOPs

3. **Profit Pumping**
    - Tool: ProfitWell (free analytics)
    - Tactic: Upsell existing clients with Breadcrumbs.

Psychological Trap #3: The "Is This All There Is?" Void

- **Fix:** Schedule "Freedom Experiments" (test working from Nomad List locations)

Case Study: From Panic To $14K/Month In 89 Days

Background:

- A 38-year-old HR director was laid off in 2024.
- $8,000 savings (Freedom Number: $15K)

Breakthrough Timeline:

- **Day 3:** Sold "AI Resume Killer" audits ($197) via LinkedIn.
- **Day17:** Automated delivery with Chatbase plus Canva
- **Day 45:** Hired $6/hr. VA to handle scheduling.
- **Day 76:** Launched $497/month "Career Immunity" membership.
- **Day 89:** Replaced corporate salary ($14.2K revenue)

Your A.I.-Powered "Business Heartbeat".

(4 Hours/Week Maintenance System)

Monday: Money & Metrics (30 min)

- Check QuickBooks + ProfitWell.
- Pay yourself first (transfer to Revolut savings vault)

Wednesday: Client Happiness (60 min)

- Send five personalized Loom updates.
- Review Typeform feedback.

Friday: Growth Engine (90 min)

- Odoo blasts to 50 new leads.
- Analyze Google Analytics traffic.

Sunday: CEO Time (60 min)

- Plan next-level moves in Notion.
- Watch one Acquired podcast episode.

Key Takeaways:

- **First 30 days** = survival on adrenaline & AI
- **Days 31-60** = systemize or suffocate
- **Days 61-90** = leverage or plateau
  A 4-hour "heartbeat" keeps you free forever

Look at your calendar. Three months from today, you could be staring at a bank statement that proves what you're capable of. Or you could be sitting at the same desk, swallowing the same excuses, waiting for permission to live. This difference doesn't revolve around fortune, talent, or even courage. It's simply a matter of who followed these steps and who didn't. You've now held a layout of 1,400 people who crossed this chasm. Their only gateway? **They kept going when the doubt screamed loudest.** The following 90 days will pass regardless. Will they find you building a legacy or just aging into regret?

Next Section: We'll design your **Digital Presence Fortress** (attract clients while you sleep).

Tonight's Task:

1. Block calendar for the next 90 days using the phases discussed.
2.   Install Wave Apps + set cash flow alerts (Your future self will thank you.

"Freedom is not achieved in a leap! It is forged in a daily decision to keep flying when the ground vanishes beneath you."

CHAPTER 10: BUILDING YOUR DIGITAL PRESENCE FORTRESS: ATTRACT CLIENTS WHILE YOU SLEEP

"In today's digital economy,
your online presence
does not just consist
of marketing, this is your
24/7 sales army."

You know that sinking feeling when you post content into the void, only to have it trickle in, yet your inbox stays silent? A gnawing sense that everyone else has cracked some secret code while you're left shouting into the digital wave. Here's the hard truth: In 2025, "build it, and they will come" is career suicide. This game has changed. Your online presence is not just a business card; this is a 24/7 sales army that works while you sleep, vacation, or finally take that midday Pilates class. Forget about vanity metrics. This is about weaponizing your digital footprint so precisely that perfect clients beg to pay you.

This section reveals:

✓   **"3-Pillar Dominance"** framework used by the top 1% creators
✓   How to engineer content that sells for you (AI-powered and effortless)
✓   The exact personal branding sequence that converts scrollers into buyers

Pillar 1: Your Content Engine (AI-Powered and Scalable)
1. "Hook → Story → Offer" Formula
(Works for all platforms)
Hook (0-3 seconds):

> •"95% of [industry] are wasting $17,000/year on this mistake."

> • **Tool:** Headline (AI hook generator)

Story (15-45 seconds):

> "I helped [client] save $23K in 3 days, here's how."
> **Tool: Claude AI** → "Write a 30-second case study about [result]."

Offer (Last 5 seconds):

"DM me 'Proof' for the exact system I used."

2. 30-Minute Content Factory:

Platform	Tool	Time/Week
Linked In	https://taplio.com/	20 Minutes
YouTube Shorts	https://www.opus.pro/	15 Minutes
Email List	https://kit.com/	10 Minutes

Pro Tip: Repurpose one core idea across seven platforms using Repurpose.io.

Pillar 2: Conversion Machine (Turn Visitors into Buyers)

1. **"No Brainer"** Landing Page:
(Must-have elements)

- **Headline:** "Stop [pain] without [common obstacle]" (Test with Headline Analyzer)
- **Social Proof:** Embed client Loom testimonials
- **CTA:** "Get [result] in [timeframe]" → Links to Calendly

Build in 17 Minutes:

1. Template: SCRIBE ($0 upfront cost)
2. AI Copy: Jasper 3
3. Trust Badges: Canva

2. DM Autopilot Sequence:

(When prospects message "How does this work?")

1. **Instant Reply:** Thanks for reaching out! Here's exactly how we help: [Loom video]."
2. **Follow-Up (24 hrs):** "Had a chance to review? I've got two spots open next week."
3. **Close:** "Shall we secure your spot?" plus Stripe link"

Tools: ManyChat (automate 90% of DMs)

Pillar 3: Authority Accelerator (Become Unignorable)

1. **"Expert Bomb"** Strategy:

- **Step 1:** Identify three niche
- podcasts. (Podchaser)

Step 2: Pitch with: "I'll share how [industry] can save $XX, XXX with AI"

- **Step 3:** Repurpose clips via Descript → Social proof.

2. **"Intel Inside"** Play:

- Partner with tools you already use: o"Official [Your

Niche] Consultant for "Software"

Example: "Click Up Certified Productivity Architect"
3.1-Hour Webinar Hack

1. Pre-record with Synthesia
2. Host "live" via StreamYard
3. Close with ThriveCart's limited offer

Case Study: From Unknown To $40K/Month In 12 Weeks

Background:

● Former insurance adjuster with 87 LinkedIn followers.

Breakthrough Sequence:
1. **Week 1-3:**
    ◻ Posted daily "Insurance AI Secrets" carousels (Canva)
    ◻ Grew to 4,200 followers with Taplio
2. **Week 4-6:**
    Launched $2,500 "Claims AI Audit" service.
    Automated delivery with Chatbase
3. **Week 7-12:**
    ◻ Partnered with Lemonade as "AI
    ◻ Consultant"

Hit $40K/month through webinar replays

Your 7-Day Digital Dominance Challenge

Day 1-2:

- Redesign LinkedIn banner with Canva + AI Generated Images.

Day 3-4:

- Film 5 Loom testimonial from beta clients Day 5-7:

- Set up ManyChat auto-responder for DMs.

Key Takeaways:

- Content is currency, and AI lets you mint it 10X faster.
- Your DM system should convert while you sleep.
- Strategic partnerships exceed going viral.

Authority is not earned: it is engineered.

Several months from now, you'll look back at your old profiles like faded Polaroids, cringing at the generic headshots, the corporate-speak bios, and posts that sounded like everyone else's because the person who implements these strategies won't be "just another expert." They'll be that expert. The one whose content gets screenshotted, whose DMs flood with ready-to-buy clients, whose name gets mentioned in industry Zooms. These resources are attainable. This framework works. The only variable left is: **Will you stay invisible, or will you finally make the internet work for you instead of against you?**

Next Section: We'll architect your **Freedom Framework** (work 20 hours/week without income drops).

Tonight's Task:

1. Redo one social profile using the 3-pillar framework.

2. Post using the Hook → Story → Offer formula

**"In a world of digital chatter, become a signal,
then watch the right people tune in."**

CHAPTER 11: FREEDOM FRAMEWORK: WORK 20 HOURS PER WEEK WITHOUT INCOME DROPS

"Our ultimate luxury is not money; reclaim your time without sacrificing income."

Honestly, you didn't leave your 9-to-5 grind to chain

yourself to a different set of working hours. That

creeping dread you feel when your phone buzzes

after dinner? That guilt when you take a midday

walk because "you should be working"? Those are

the ghosts of your old corporate life haunting

your new freedom. But here's the liberating truth:

Building wealth no longer requires trading time for

money. Your most precious assets aren't your skills

or even your clients. What are you diverting your

attention to?

✓ "4-Hour CEO" system used by 7-figure solopreneurs.
✓ How to automate client fulfillment. (AI plus human hybrid model)
✓ Your "freedom metrics" dashboard. (track what matters)

Part 1: 4-Hour Ceo Workweek Design

1. Daily 20-20-20 Method

(60 minutes/day that drives 90% of results)

Time	Focus Area	Tools
20 min	Revenue Generation	Adobe Firefly (AI video outreach)
20 min	Client Happiness	Loom (Async updates)
20 min	System Optimization	Zapier (Fix one automation)

Pro Tip: Use Focusmate to stay accountable.

2. "CEO Friday" Schedule:

(3 Non-Negotiable Tasks Weekly)

1. **Profit Check:** QuickBooks plus ProfitWell
2. **Team Sync:** 15-min Loom update for VAs
3. **Offer Audit:** Test one price increase with Breadcrumbs

Part 2: A. I. Human Hybrid Fulfillment Model80/20 Outsourcing Matrix:

Task Type	Who/What Handles It	Cost
Repetitive (80%)	AI Tools	$20-$100/month
Creative (15%)	Overseas VA	$3-$8/hour
Strategic (5%)	YOU	Priceless

Sample Workflow:

1. Client signs up → Typeform triggers Zapier
2. Durable AI generates the first draft deliverable.
3. $5/hr. VA from OnlineJobs.ph adds final touches.
4. You review the final product. (3 min)

Part 3: Your Freedom Metrics Dashboard

(Track These Instead of Vanity Metrics)

Metric	Tool	Target
Revenue/Hour	Clockify	$500+
Client Happiness	Delighted	9/10 NPS
System Health	Notion	All automation Green

Golden Rule: If a metric can't be improved in 20 minutes, stop tracking it.

Case Study: From Burnout To 4-Hour Work Weeks

Background:

- Former consultant working 60-hour weeks for $15K/month.

Transformation:

1. **Month 1:** Automated client reports with Synthesia
2. **Month 2:** Hired $4/hr VA for admin via Upwork
3. **Month 3:** Implemented "CEO Friday" procedures.
4. **Now:** Earns $22K/month working 10-15 hours weekly.

Your 7-Day Freedom Challenge

Day 1-2:
- Install Clockify and track all work hours.

Day 3-5:
- Identify one task to offload to AI (Refer to Chapter 7 tools)

Day 6-7:
- Schedule the first "CEO Friday" in the calendar

Key Takeaways:

- Time freedom begins with ruthless prioritization.
- AI handles execution, and you handle direction.

What gets measured gets optimized?

Next Section: We'll explore **Scaling Without Limitations** (growing beyond yourself).

Tonight's Task:

1. Delete three unproductive metrics from your tracking.
2. Block 60 min/day for the 20-20-20 method.

Reflect: It's sunrise on an ordinary Monday, and you're sipping tea while the world wakes up to the daily grind; you wake up to the quiet certainty that your business runs itself. Your AI team handles client requests, your systems generate income, and your VA keeps everything flowing. This is possible; it's already occurring for those who have hope to believe in a better way. If you take just one initial step, trust that you will be compensated for your boldness. Your future self is already cheering you on.

"Faith does not wait for a perfect moment; it helps us build our wings on our way up."

CHAPTER 12: SCALING WITHOUT CHAINS: HOW TO GROW BEYOND YOURSELF WITHOUT BECOMING A BOSS

"An ultimate lie about scaling? That you need employees, offices, or investor money. New empire builders leverage AI and strategy. partnerships instead."

Y ou've felt it, that subtle panic when you realize scaling your business the old way means recreating corporate chaos you escaped. Hiring employees feels like adopting needy children. Office leases chain you to a location. Investor demands to hijack your vision. But what if growth didn't require sacrifice? What if you could 3X your revenue without tripling your workload or compromising your soul? Modern-day underdogs have found a way to scale differently, utilizing AI as their workforce, partnerships as their leverage, and freedom as their non-negotiable key performance indicator (KPI). That era of "more overhead = more success" has diminished. Welcome to scaling without restraint.

✓ "Asset-Light Scaling" framework (3X revenue without 3X work)

✓ How to build a $100K/year "ghost workforce" for < $1K/month

✓ A partnership playbook that replaces traditional hiring

Part 1: Asset-Light Scaling Pyramid

1. AI s Your First Employees (Layer 1)

Role	AI Tool	Monthly Cost
Customer Service	Chatbase	$19
Content Team	Jasper + Suno	$99
Operations Manager	Zapier + Make	$50

Total AI Workforce Cost: $168/month (vs. $15K+ for humans)

2. Fractional Talent Cloud (Layer 2)

- **$3-$15/hour specialists:**
  Upwork (project-based)
  Fiverr Pro (pre-
  vetted)
  Toptal
  (premium)

- **Key Rule:** Hire for **outcomes**, not hours.

3. Strategic Partnerships (Layer 3)

- **Revenue-Sharing Models:**
  ◦"Bring your own tool" discounts
  (PartnerStack)

o   Co-branded offers with complementary businesses.

Part 2: $100K Ghost Workforce Design

Step 1: Productize Your Knowledge

- Turn your process into:
  - Digital courses :(Kajabi)
  - SAAS Tool: (Bubble IO)
  - Certification Program: (Thinkific)

Step 2: Automate Fulfillment

- **Client Onboarding:**
- Typeform → Zapier → Notion

Delivery: Synthesia - AI videos + Convert Kit email sequences.

Step 3: Scale Acquisition

- **Paid Ads:** AdCreative.ai + Writesonic
- **Organic:** Taplio LinkedIn automation

Part 3: Partnership Playbook

1. **"Intel Inside" Strategy**:

    • Become the **official [your expertise] partner** for the software your clients already use.

    • Example: "Certified Click Up Consultant."

2. **Revenue Share Alliance:**

    * Partner with non-competing service providers
    *
    Tool: PartnerStack (manage all partnerships)

3. **Affiliate Army**

    * Recruit micro-influencers with UpPromote.
    *
    Pay 30-50% commissions for life.

Case Study: $0 To $83K/Month With 0 Employees

Background:

- Former real estate agent with no tech background.

Growth Sequence:

1. **Month 1-3:**
   - Built an AI-powered "Property Analyzer" tool. (Bubble)
   - Automated reports with Synthesia.

2. **Month 4-6:**
   - Partnered with 12 brokerages (PartnerStack)
   - Hired $6/hr VA to handle onboarding.

3. **Month 7-12:**
   - Scaled to 1,200+ users.
   - Revenue: $83K/month (90% profit margins)

Your 30-Day Scaling Challenge

Week 1:

- Identify one process to produce.

Week 2:

- Set up the first revenue-sharing partnership.

Week 3-4:

- Launch micro-affiliate program.

Key Takeaways:

- AI handles volume, and humans handle exceptions.
- Partners > Employees for asset-light growth

Productization is the ultimate leverage.

Next Section: We'll define '**A New American Dream**' (legacy beyond money).'

Tonight's Task:

1. List of 3 potential partners in your niche
2. Research one no-code tool to productize your knowledge.

Doubt will creep in: **This is too simple. Where's that grind? Enduring through Trials?** But look around; proof is everywhere. Consultant serving Fortune 500 companies from a beach in Bali. That single mom outsells corporations with just AI and a prayer. They're not special. They refused to believe growth requires bondage. As you stand at this intersection, remember every business builder before you faced this exact moment of choice between conventional "success" and genuine sovereignty. All that's missing is your faith in taking the first step into this new paradigm, where less really does become more.

> **"Our mightiest oaks grow from single seeds, not tangled roots."**

CHAPTER 13: NEW DREAM: BUILDING A LEGACY BEYOND A MATRIX

"An old dream was
retirement at 65.
The new dream is freedom at 35
and a legacy that outlives you."

"What if your freedom could become someone else's hope?"

That old American Dream was a solitary climb, hoard, and retire alone. The new dream breeds something more tangible: Think of your AI tools as not only automating your income but also training single parents in your community to do the same. This does not solely consist of building a lifeboat for one; will you be the one who crafts a lighthouse? Because when you redefine success, you don't just change your life... You set a path for others to create.

This final section reveals:

✓ 3 Pillars of True Wealth (money is just 1/3)
✓ How to engineer generational impact without Wall Street or Silicon Valley.
✓ Your "post-Matrix" compass (what to do when you're finally free.)

Pillar 1: Freedom Trifecta:

1. Time Sovereignty:

   - **Metric:** % of calendar YOU control
   - **Tool:** Reclaim.ai (AI time blocking)
   - **Target:** 90% discretionary time.

2. Location Liberation:

   - **Strategy:**
   - Build "geo-arbitrage" income streams.

     Tools:
     ◦ Nomad List (find low-cost hubs)
     ◦

Wise (borderless banking)

3. Purpose Capital:

- **Exercise:** "If money disappearswhat would I still do daily?

  Framework: Ikigai model Resource Tools - (lkigai Tribe)

Pillar 2: Generational Wealth Engine1. "Infinite Asset" Framework

Asset Type	Examples	Tools To Build
Digital IP	Courses, software content	Kajabi
AI Systems	Chatbots, automation	Make.com
Community	Paid memberships	Circle

2. 1% Legacy Rule

*Invest 1% of weekly time/money into:

a) Mentoring others (ADPList)

b) Building "fun" projects (Mobirise)

Pillar 3: Post-Matrix Compass

When You Wake Up Unbound...

1. **First 90 Days:** Explore, create, and detox from the corporate mindset.
2. **Months 4-12:** Double down on what brings joy + profit.
3. **Year 2+:** Seed "legacy projects" with 10% of income.

Warning Signs You're Still Plugged In:

- Checking email constantly "just in case"
- Feeling guilty about leisure time
- Identifying by past job title

Case Study: From Burnout To Multi-Generational Impact Background:

- Former finance director (80-hour weeks, $250K salary)

Transformation:

1. **Year 1:** Built a $40K/month AI consulting biz.
2. **Year 3:** Launched a free coding school for

  foster kids.

3. **Year 5:** 73 students placed in tech jobs ($6M + collective earnings)

"My W-2 career paid for bills. My freedom builds legacy."

Your Legacy Layout Challenge:

- **Month 1:** Define your "freedom trifecta" metrics.

- **Month 2:** Launch one "infinite asset"

- **Month 3:** Commit to a 1% legacy rule.

Final Key Takeaways:

- True wealth = Time + Freedom + Purpose
- Legacies are built with systems, not just savings. The Matrix's final trick is making you fear life outside it.

Last Page Of An Old Chapter (Your New Beginning)

"You've now got a map to assist in your navigation. Only question left:

Will you spend the rest of
your life wondering what
it could have been, or walk
through this door you
have just discovered?"

"Your breakthrough doesn't just involve you; create a spark that lights a thousand fires in the dark."

A NOTE TO YOUR FUTURE SELF

(Read when you doubt, when you celebrate, and when you need to remember why you began.)

Dearest Innovator,

I sense your struggle; I can relate to you standing in that doorway between a life you were told to live and the one you're brave enough to build. This air smells different here, doesn't it? Less like stale office coffee and fluorescent lights, more like the possibility of a life where you are free to be creative.

There will be days when the old fears creep in. Days when a LinkedIn post about someone's **"VP promotion"** will make your stomach clench. Days when numbers in your bank account tempt you to **"go back."** On those days, I want you to do something radical: **Pull up the photos from your last corporate job.** Zoom in on your eyes. That moment of exhaustion, that dull resignation, that's what you're running from. Now look in the mirror. However messy this freedom feels, your eyes are alive again.

This decision was not just about finances. It was about reclaiming your mornings. About school plays attended in broad daylight, about the right to say **"no"** without panic. You've already overcome a great battle: You chose yourself. The rest is just fine-tuning.

When the doubt comes (and it will), remember:

> **1.**The first client who paid you real money for your genius.
> **2.**The first time, you worked from a beach, park, or cozy café and thought, "I can't believe this is my life."
> **3.** The people watching you, emboldened by your courage, are now dreaming bigger because you have proved it's possible.

You are not just building a business. You're creating a layout for those still trapped in their self-imposed Matrix.

With love,

You Who Believed First

PS: That sting you're stressing over right now? It is already worked out. Breathe!

IDENTIFIER
#979-089998970-0-8